Math Concept Reader

The World's TALLEST Buildings

By Ilse Ortabasi

Copyright © Gareth Stevens, Inc. All rights reserved.

Developed for Harcourt, Inc., by Gareth Stevens, Inc. This edition published by Harcourt, Inc., by agreement with Gareth Stevens, Inc. No part of this publication may be reproduced or transmitted in any form or by any means, electronic or mechanical, including photocopy, recording, or any information storage and retrieval system, without permission in writing from the copyright holder.

Requests for permission to make copies of any part of the work should be addressed to Permissions Department, Gareth Stevens, Inc., 330 West Olive Street, Suite 100, Milwaukee, Wisconsin 53212. Fax: 414-332-3567.

HARCOURT and the Harcourt Logo are trademarks of Harcourt, Inc., registered in the United States of America and/or other jurisdictions.

Printed in China

ISBN 13: 978-0-15-360195-8
ISBN 10: 0-15-360195-7

6 7 8 9 10 0940 16 15 14 13 12 11 10 09

Chapter 1:
Brainstrorming

The students in Mr. Cowaltowski's class are beginning a project for the World Geography Fair. Mr. C., as his students call him, looks forward to this project each year. His classes have created many interesting projects throughout the years. This year, his class will study some of the world's tallest buildings.

The class is excited. They have never researched the world's tallest buildings and look forward to learning about them. They will compare the heights of these buildings and find their locations on a map. They also will look at population data for the cities where these buildings are located. The students will conduct research and create an original display to show their data. They will use both math skills and geography skills to complete their project.

Mr. C. tells his students that their project will be displayed in the school lobby in a few weeks. There is a lot of work to do. It will take a lot of teamwork to make this project successful.

Mr. C. encourages the students to share their ideas. He asks everyone to come up with ideas on how to display the completed project.

Caitlin suggests creating scaled models of the world's tallest buildings. Errol thinks that a multimedia display would be interesting. He thinks they could show photographs and videos of the tallest buildings. Hassan suggests they create graphs to compare the sizes of the world's tallest buildings. Will thinks that the class could make silhouettes of the buildings.

"You have very good ideas," says Mr. C. "How could we combine these ideas for the class project?" The class buzzes with excitement. The students discuss the different ideas among each other and finally decide to make a mural that shows drawings of the buildings. The background of the mural will be a world map. They will add data about the heights of the buildings and the population from the cities where the buildings are located.

Mr. C. and all the students like this idea. None of his classes have ever created a mural for the World Geography Fair.

Mr. C.'s students will create a mural of the world's tallest buildings for the World Geography Fair.

The Two International Finance Centre stands tall against the Hong Kong, China, skyline.

Mr. C. looks at his list. He has the names of the seven tallest buildings in the world. He divides the class into seven groups with four students in each group. He assigns each group a tall building to research. The assignments include two buildings in the United States: the Empire State Building in New York and the Sears Tower in Chicago.

The other tall buildings on the list are located on the Asian continent. In fact, three buildings are found in China, Jin Mao Tower, Shanghai, China; CITIC Plaza, Guangzhou, China; and Two International Finance Centre, Hong Kong, China. One of the buildings, Taipei 101, is located in Taipei, Taiwan. Finally, there are twin towers, Petronas Towers 1 and 2, which are located in Kuala Lumpur, Malaysia.

Mr. C. remarks that there are many more tall buildings planned or under construction all over the world. Most of them will be taller than the buildings already on the list. The students look at the world map to find the cities where the buildings are located. The class is off to a great start!

Mr. C. knows the class has a lot of work to do to finish the project in time for the World Geography Fair. He schedules time in the media center so the students can search for the information they need. The media specialist will help them to use reference books, maps in an atlas, and the Internet. The students will make sure to share any good resources they find in the media center with each other.

Each group plans their part of the class project for the World Geography Fair. Alex wonders, "Are the world's tallest buildings located in cities that have larger populations?" "Well," says Katie. "There must be workers to fill these tall buildings." The class decides to investigate.

Two students in each group will look for information about the building assigned to their group. They will gather data on the height of that building.

The other two students in each group will find population data on the city where their building is located. The class will compare their data. They will compare the heights of some of the world's tallest buildings as well as the population in the cities where these tall buildings are located.

Using the Internet, these students research the world's tallest buildings and the populations of the cities where the buildings are located.

Chapter 2:
Comparing the Data

The students return from the media center and Mr. C. asks each group to share the data they collected. Madeleine says, "I read that the height of a building is measured from the sidewalk level to the top of the building." "That's right," says Mr. C. "The height does not include antennas or flag poles on top of the building, but it does include spires. The Taipei 101 building has a 60-foot spire adding to its height."

Most groups found data about the heights of the buildings in meters and in feet. The class agrees to compare the height of the buildings in feet and each group rounds the height to the nearest whole number.

Mr. C. creates a table to record the data. He reminds the class that it is helpful to stay organized and accurate when recording data. "Using a table will make it easer to order and compare the data you have found."

Each group reports on the height of a building. When they are finished, the data table looks like this:

Building, City, and Country	Height (in Feet)
Empire State Building, New York, USA	1,250
Sears Tower, Chicago, USA	1,451
Jin Mao Tower, Shanghai, China	1,381
CITIC Plaza, Guangzhou, China	1,283
Two International Finance Centre, Hong Kong, China	1,362
Taipei 101, Taipei, Taiwan	1,667
Petronas Towers 1 and 2, Kuala Lumpur, Malaysia	1,483

Now that the class has recorded the heights of the buildings, Mr. C. suggests that they order the buildings from tallest to shortest. "We can do this task together," says Mr. C.

Melissa has an idea. "We need to look at place values of the numbers to compare them," she says. "Great idea," Alex replies. "We compare the digits of all the height measurements in feet, starting with the digits on the left." Naomi joins in. "Yes," she says. "The height of each building is a 4-digit number."

Mr. C. says, "The number in the thousands place for all the building heights is 1. We'll need to look at the digit in the hundreds place for each building's height to compare the heights." "Look at the number 1,667, the height of Taipei 101," says Caitlin. "It has the greatest number of hundreds of all the building heights."

"That's right," says Errol. "That means Taipei 101 is the world's tallest building." Mr. C. writes Taipei 101 at the top of the list.

At 1,667 feet, the Taipei 101 Building in Taipei, Taiwan, is the tallest building on the list.

Now Mr. C. says, "Next, look at the data on building heights and find the buildings with a four in the hundreds place." The students carefully study the table Mr. C. has created. There are two buildings, the Petronas Towers in Kuala Lumpur, Malaysia and the Sears Tower in Chicago, USA.

Mr. C. says, "We want to find out which building is the tallest so we'll need to order these numbers by looking at the tens place." Sam and Katie look at the data for the Petronas Towers. The towers are 1,483 feet tall. The number in the tens place is eight. They look at the data for the Sears Tower. The Sears Tower is 1,451 feet tall. The number in the tens place is 5. Katie says, "Five is less than eight, so the Sears Tower is shorter than the Petronas Towers."

Lee agrees. "I can see that the Petronas Towers come next in height. The Petronas Towers are just 32 feet taller than the Sears Tower. I read that the Petronas Towers have 88 floors inside them!"

The paired Petronas Towers are an eye-catching sight in Kuala Lumpur, Malaysia.

The class continues to compare the heights of the buildings. They order the data table and list the buildings from tallest to shortest. "It will be fun to keep track of new buildings being built around the world as they are finished," says Hassan. "That way, we can see if the world's tallest building has changed over time."

The data is complete and the comparisons are done. The students look at the data table. This is what it looks like:

Building, City, and Country	Height (in Feet)
Taipei 101, Taipei, Taiwan	1,667
Petronas Towers 1 and 2, Kuala Lumpur, Malaysia	1,483
Sears Tower, Chicago, USA	1,451
Jin Mao Tower, Shanghai, China	1,381
Two International Finance Centre, Hong Kong, China	1,362
CITIC Plaza, Guangzhou, China	1,283
Empire State Building, New York, USA	1,250

Mr. C. reminds the class of the population research question that Alex asked earlier. "Based on your research, are the world's tallest buildings located in cities that have larger populations?" Taipei 101 is the world's tallest building. It has 101 floors. Will wonders if Taipei, Taiwan, is the city with the largest population among the seven cities where these tall buildings are located.

Alex suggests that they compare the population data using a similar approach to the one they used for comparing the heights of the buildings. They will list each city where the buildings are located and then order the population data from the same year.

Katie says, "I have been to Chicago many times because my grandparents live there. It is a big city. I think that it might have the largest population of all the cities where the tall buildings are."

The students look at the population data for 2005 that they collected during their research in the media center. Mr. C. lists the seven cities where the world's tallest buildings are located. They include Chicago and New York City in the United States; Shanghai, Guangzhou, and Hong Kong in China; Kuala Lumpur, Malaysia, and Taipei, Taiwan.

Mr. C. asks each group to report the population data they found for each city and then he records the data in a spreadsheet. He projects the data for the class to view and discuss together.

Once again, the students decide to compare the data by listing the cities from the largest population to the smallest population. They have already had practice doing this by listing the buildings in order of their heights. Melissa reminds them, "We need to look at place values of the numbers to compare them."

City Name	Population
Chicago, USA	8,620,000
New York, USA	20,160,000
Shanghai, China	10,000,000
Guangzhou, China	4,500,000
Hong Kong, China	6,720,000
Kuala Lumpur, Malaysia	4,760,000
Taipei, Taiwan	5,730,000

The students look at the data and see that New York City has the largest population of the seven cities with the world's tallest buildings. The Empire State Building, however, is not the world's tallest building.

Sam is curious. "I wonder what the total population is of the two cities with the tallest buildings in the United States." Katie suggests they add the population of New York City and Chicago to find out. The students look at the table. New York City has a population of 20,160,000. Chicago has a population of 8,620,000.

20,160,000 + 8,620,000 = 28,780,000

More than 28 million people live in these two U.S. cities.

Next, they decide to find out the population of the cities in China with the world's tallest buildings. They add the population of Shanghai to the population of Guangzhou and Hong Kong. Their equation looks like this:

10,000,000 + 4,500,000 + 6,720,000 = 21,220,000

More than 21 million people live in these three cities in China.

About 4.5 million people live in Guangzhou, China, the location of the CITIC Plaza.

Errol looks at the population data again. He wants to find the difference between the city with the largest population and the city with the smallest population.

"The city with the largest population is easy to find," says Will. "It's New York City. There are 20,160,000 people who reside in New York City." Katie adds, "The city with the smallest population is either Kuala Lumpur or Guangzhou."

The students look at the population data. Will says, "To find out which of these two cities has the smallest population, we'll need to order these numbers by looking at the hundred thousands place." Errol and Katie look at the data. Kuala Lumpur, Malaysia has a population of 4,760,000 people and Guangzhou, China has a population of 4,500,000 people. For Kuala Lumpur, the number in the hundred thousands place is 7. For Guangzhou, the number in the hundred thousands place is 5. "Since five is less than seven, Guangzhou is the city with the smallest population out of the seven cities we are studying," says Will.

City Name	Population
New York, USA	20,160,000
Shanghai, China	10,000,000
Chicago, USA	8,620,000
Hong Kong, China	6,720,000
Taipei, Taiwan	5,730,000
Kuala Lumpur, Malaysia	4,760,000
Guangzhou, China	4,500,000

Now Errol compares the population data from New York City and Guangzhou. The answer will tell him the difference between the city with the largest population and the city with the smallest population. Errol subtracts.

20,160,000 − 4,500,000 = 15,660,000

New York City has 15,660,000 more people than Guangzhou. Caitlin says, "Look at the data. Of all seven cities, the city with the shortest building actually has the greatest population." Mr. C. smiles. "That's a good observation," he says. The class is surprised. They did not expect the city with the greatest population to have the shortest building.

Finally, the students have completed the data tables. They have compared the heights of the world's tallest buildings as well as the population in the cities where those buildings are located. The data collection has been interesting. The class has discovered that the cities where the world's tallest buildings are located all have populations over four million people. Alex concludes that the tallest building in the world is not in the city with the largest population among these seven cities.

More than 15 million people live in New York, home of the Empire State Building.

Chapter 3:
Tallest Buildings on Display

It's now time for the class to begin putting their mural together for the World Geography Fair. Mr. C. places paper, markers, and rulers on the table in the back of the classroom. He adds colored pencils and paint, too.

"You have the data about the heights of the buildings and the population from the cities where the buildings are located," says Mr. C. "Now it's time to plan the mural."

The groups begin the mural for the World Geography Fair. Each group makes a large drawing of their building and places it on a background of the world map. They place it near the city and country where it is actually located. They include a title and the names of the countries and cities where the buildings are located. They include the data table for the heights of the buildings and the population data for the cities in which the buildings are located. Some students use computers to make data tables. They work all afternoon until the mural is complete.

This student uses the computer to work with data for the mural.

Finally the big day arrives. The class works all morning in the school lobby to prepare and assemble their mural. They mount their pieces for the World Geography Fair display and place them where they can easily be seen by visitors. Mr. C. reminds them to include a title, the names of the countries and cities where the buildings are located, the data table of the heights of all the buildings and the population data for each city. Finally, they add a list of references where the data were found.

The World Geography Fair display looks great in the lobby. The mural is interesting to look at and informative. Each group did a wonderful job and the students can hardly wait for their families to see it. Mr. C. is eager to tell the parents how pleased he is about how well the class worked together. He is sure that the parents will admire the mural. He also thinks they will learn something from it. It will be the first thing the parents see when they come in the door! The students are proud of their work.

Students work on part of the mural for the World Geography Fair.

Glossary

antenna a device to receive or send radio signals

brainstorming an activity in which members of a group contribute ideas, often to solve a problem

digit any one of the ten symbols 0, 1, 2, 3, 4, 5, 6, 7, 8, or 9 used to write numbers

geography the science that describes the features of Earth's surface

height the length of a perpendicular from the base to the top of a plane figure or solid figure

hundreds place the position of the third digit from the left of the decimal place in a number

model a smaller version representing something that is larger

multimedia using or involving several media such as video, music, and models

mural a painting on a wall or ceiling surface

place value the value of a digit as determined by its position in a number

population all the people living in a country, city, or area

spire tapering, pointed structure on the roof of a tower or building

Photo Credits: cover, title page, p. 13 © Jose Fuste Raga/Corbis; p. 3 © Jim Cummins/Corbis; p. 4 © Mike Kemp/Corbis; p. 5 © Gabe Palmer/Corbis; p. 7 © Simon Kwong/Reuters/Corbis; p. 8 Tengku Bahar/AFP/Getty Images; p. 11 © David Butow/Corbis Saba; p. 14 © Corbis; p. 15 © Richard Hustings/Photo Edit.